The Library of Subatomic Particles
The Neutron

Fred Bortz

The Rosen Publishing Group, Inc., New York

To Susan, who long ago set off a chain reaction for me

Published in 2004 by The Rosen Publishing Group, Inc.
29 East 21st Street, New York, N.Y. 10010

Copyright © 2004 by The Rosen Publishing Group, Inc.

First Edition

All rights reserved. No part of this book may be reproduced in any form without permission in writing from the publisher, except by a reviewer.

Library of Congress Cataloging-in-Publication Data

Bortz, Alfred B.
The neutron / by Fred Bortz.— 1st ed.
   p. cm. — (The library of subatomic particles)
Summary: A look into the discovery of the neutron, which completed our picture of the structure of the atom and enabled us to explain the existence of isotopes and understand how nuclear fission occurs.
Includes bibliographical references and index.
ISBN: 978-1-4358-3662-4
1. Neutrons—Juvenile literature. 2. Particles (Nuclear physics)—Juvenile literature. [1. Neutrons. 2. Particles (Nuclear physics)] I. Title. II. Series.
QC793.5.N462B67 2004
539.7'213—dc22

2003013096

Manufactured in the United States of America

On the cover: an artist's illustration of the exchange particles called gluons holding protons and neutrons together inside an atomic nucleus

Contents

	Introduction	4
Chapter One	The Atomic Nature of Matter	6
Chapter Two	Particles Inside the Atom	17
Chapter Three	The Nucleus, Neutrons, and Nuclear Forces	28
Chapter Four	Neutrons and Nuclear Fission	41
	Glossary	53
	For More Information	58
	For Further Reading	60
	Bibliography	61
	Index	62

Introduction

About two centuries ago—in 1803 to be precise—British chemist John Dalton transformed science with a simple statement: All matter is made of atoms. The idea was far from a new one. The ancient Greek philosophers Leucippus and Democritus had suggested it twenty-three centuries earlier when they asked themselves how small they could cut a piece of matter without changing its nature. They described the smallest possible piece as *atomos*, meaning "indivisible." These were only ideas, however, while Dalton connected the notion of atoms to laboratory observations.

In Dalton's theory, the smallest pieces of most substances turn out not to be atoms but molecules. A water molecule is made of two hydrogen atoms and one oxygen atom. Thus it is not indivisible, but it is still the smallest speck of matter that can still be called water. Dalton spoke of two kinds of substances: elements, which are made of only one

Introduction

kind of atom; and compounds, which are made of only one kind of molecule.

Dalton, like Leucippus and Democritus, thought that atoms were indivisible, but that turns out to be incorrect. As scientists discovered more elements, they began to wonder what distinguished one element from another. Might atoms be made of even smaller particles, just as molecules are made from atoms? By the end of the nineteenth century, physicists (scientists who study matter and energy) had begun to find evidence that the answer was yes.

Today, students like you know the names of the most important of the subatomic particles. You have probably learned that atoms contain protons, neutrons, and electrons, but you may not know how those particles were discovered and how they combine to give atoms their properties. You may not know that those three are only part of the full set of particles in the universe. This book will carry you deep inside the atom, where you will meet the neutron and discover the great power that it can unleash.

Chapter One

The Atomic Nature of Matter

Those ancient Greeks began with two simple questions: What is matter made of, and why do different kinds of matter behave differently? When they proposed that matter is made of atoms, they also suggested that the properties of a particular kind of matter depended on the shape and texture of its atoms. For instance, water atoms would be round and smooth, while rock would be composed of atoms that are hard and sharp or gritty.

However, they never considered testing their atomic theory by observing nature. It was not because atoms were too small to see (although they are), but rather because testing ideas through observation, a cornerstone of modern science, had not yet become part of human culture. In fact, Greek philosophy took an opposite approach, valuing pure human thought

The Atomic Nature of Matter

above all else. Logical thinking and reason alone were considered sufficient to deduce the truth. Great thinkers such as Socrates, Aristotle, and Plato used their powerful minds and logic to deduce what they believed to be the truth about the world around them.

Thoughts Matter. The ancient Greek philosopher Aristotle was considered to be such a brilliant thinker that his ideas about the nature of matter were not challenged for nearly 2,000 years.

Aristotle was considered so brilliant that people would rarely question his ideas, even long after his death. For nearly 2,000 years, people simply accepted Aristotle's explanation that all the world's matter was made of four elements: earth, air, fire, and water. The idea of atoms all but disappeared. Today, we know that both Democritus and Aristotle were right in one way but wrong in another.

There is a limit to how small a piece of matter can be cut and still remain the same

The Neutron

substance, so Leucippus and Democritus were right about that. But since most substances are compounds, the smallest possible piece is a molecule instead of an atom. Furthermore, that piece is not indivisible. Molecules of compounds can be divided into atoms, which contain even smaller (subatomic) particles. Aristotle's concept of elements was correct, but not the ones he wrote about. The number of natural elements is nearly 100, and Aristotle's four are not among them. Water is a compound of hydrogen and oxygen. Both earth and air are mixtures containing both elements and compounds. And fire is not matter at all but energy produced by a rapid chemical reaction.

Between ancient Greek natural philosophy and today's atom-based chemistry came a practice called alchemy, in which people tried to make certain substances out of others. Most often, alchemists were searching for ways to turn less valuable metals into gold. We now know that the techniques they used—which often produced real chemical changes—were doomed to failure. Both ancient alchemy and modern chemistry can rearrange the way atoms

The Atomic Nature of Matter

are combined, but they can't change one kind of atom (such as lead) into another (like gold).

Many of the most famous alchemists were frauds, but others developed a rudimentary knowledge of matter and techniques that they used to extract or purify many useful elements and compounds from natural minerals and ores. By the beginning of the eighteenth century, alchemy had become chemistry. Chemists of that time investigated many important phenomena in their laboratories: the behavior of gases at different pressures and temperatures; processes of combustion and corrosion; and the relationship between electricity and matter. None of these were fully understood, but the chemists collected a great deal of valuable information through systematic scientific measurement and study.

At the turn of the nineteenth century, English scientist John Dalton (1766–1844) was turning his attention from meteorology to chemistry, expecting that he would understand the weather better if he knew more about the gases of the air. He quickly figured out that the old idea of atoms would explain many properties of gases

The Neutron

It's Elementary. In 1810, John Dalton revolutionized chemistry with his atomic theory of matter. He divided substances into elements, which are composed of only one kind of atom, and compounds, which are composed of atoms in particular combinations called molecules.

and chemical processes. In 1810, Dalton published *A New System of Chemical Philosophy*, a book that revolutionized chemistry. Dalton began with the statement that all matter is made of atoms. He went on to explain that each element is made of a particular kind of atom, that all of its atoms are identical to each

The Atomic Nature of Matter

other, and that atoms join together to form compounds—always in small whole numbers, no fractions allowed. The properties of atoms, such as weight, distinguish one element from another.

Applying those simple rules, he determined the atomic weight of different elements. He set the atomic weight of hydrogen, the lightest element, at one unit, and he calculated the atomic weight of other atoms from that. He wasn't always correct, but it was clear he was making progress. For instance, he knew that water had eight times as much oxygen as hydrogen by weight. But he mistakenly assumed that a water molecule had one atom of each element, and therefore he concluded that the atomic weight of oxygen was eight. When later research showed that water molecules had two atoms of hydrogen for each oxygen atom, scientists corrected the atomic weight of oxygen to sixteen.

Classifying the Elements

Dalton had given chemistry a new basic vocabulary. Scientists everywhere spoke of

The Neutron

elements and compounds, atoms and molecules, and, of course, they asked questions. How many elements are there, they wondered, and how could they classify the growing number? Several properties provided hints of similarities and patterns among the elements. These included melting or boiling points, densities (how much each cubic centimeter weighs), the way one element combines with others, and the properties of the resulting compounds. Still, no one had successfully turned those hints into a classification scheme. Then, in 1869, Russian chemist Dmitry Ivanovich Mendeleyev (1834–1907), a professor at St. Petersburg University, had a flash of insight.

Mendeleyev was a walking encyclopedia of the sixty-three known elements, with detailed knowledge of their properties. He decided to make a set of cards, one for each element, listing the known properties of each, and arranged them in order of increasing atomic weight. For three days and nights, he lay the cards on his table, grouping and regrouping them incessantly, hoping to discover a classification scheme before

The Atomic Nature of Matter

Order, Please. By 1869, the number of known elements had grown to sixty-three, and scientists had noticed some similarities in their properties. But no one could lay out those similarities in an orderly arrangement until St. Petersburg University chemistry professor Dmitry Ivanovich Mendeleyev envisioned the periodic table of the elements.

leaving on a long-scheduled train trip to the countryside, where his family owned an estate. Just before the time came for him to leave, the weary professor fell asleep and dreamed of playing solitaire with his deck of element cards. Awake on the train, Mendeleyev played element solitaire, and by

The Neutron

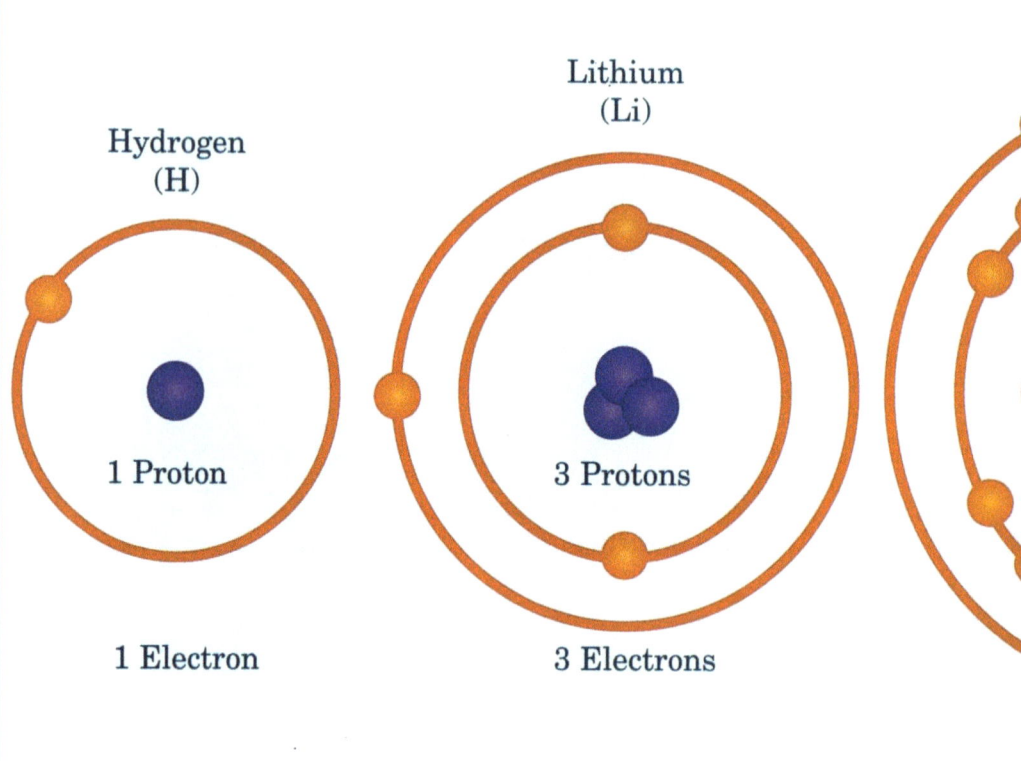

Valence. The columns of the periodic table consist of elements with the same chemical property known as valence. Elements with the same valence form similar compounds. For instance, hydrogen and the alkali metals (such as lithium and sodium shown here) make up column I of the periodic table, and all form compounds with oxygen in the ratio of two atoms to one, like the familiar chemical formula of water, H_2O. It took scientists more than fifty years after the discovery of the periodic table to understand that valence was related to the electrons left over after other electrons filled "shells" around a nucleus.

The Atomic Nature of Matter

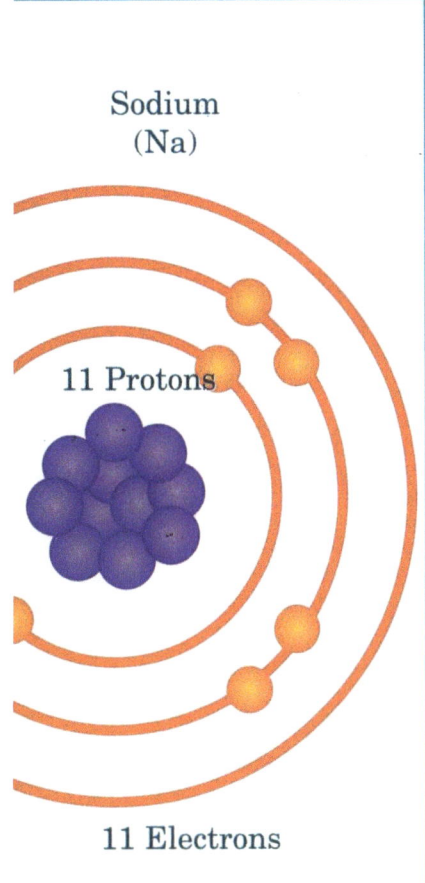

Sodium (Na)
11 Protons
11 Electrons

the time he arrived at his destination, an arrangement of elements in rows and columns had begun to take shape. With the elements ordered by increasing atomic weight down the columns from top to bottom, he discovered that the horizontal rows of elements aligned themselves to match a chemical property known as valence. That property relates to the numbers of atoms of two combining elements. For example, the alkali metals—lithium, sodium, potassium, rubidium, and cesium—which all have a valence of +1, fell into alignment across one row. Likewise, the nonmetals

The Neutron

called halogens—fluorine, chlorine, bromine, and iodine—with a valence of –1, lined up in another row. Because of its repeating pattern, Mendeleyev called this arrangement the periodic table of the elements. (Today, the standard arrangement of the periodic table is increasing atomic masses across rows, with the valences aligned in columns.)

The table had a few gaps, which Mendeleyev boldly claimed would be filled by undiscovered elements that fit the pattern. He was right, down to the atomic weights, densities, and other properties of those elements! But he couldn't have known that the quest to understand those atomic weights would lead another famous scientist to predict another undiscovered bit of matter even smaller than an atom—the neutron.

Chapter Two

Particles Inside the Atom

As more elements were discovered and the periodic table became more important to science, two new questions arose. What makes each element behave as it does, and what makes the periodic table periodic? The answer to both begins with this fact: Despite the meaning of the Greek word from which they get their name, atoms are not indivisible. They contain subatomic particles.

The first subatomic particle to be discovered was the electron. J. J. Thomson (1856–1940) of Cambridge University in England was studying an interesting electrical phenomenon in glass tubes from which most of the air had been removed. When scientists inserted pairs of electrodes into the tubes and connected them to the opposite ends of a source of electricity, like a battery, they observed a glow near the negative electrode, or cathode. They didn't

The Neutron

know what was causing the light, but they called it a cathode ray. Some scientists thought the cathode rays were probably waves of energy; others said they were streams of particles. In 1897, Thomson demonstrated that cathode rays were streams of the smallest particles ever known. He called them corpuscles, but today we call them electrons. By his measurements, a single corpuscle had less than a thousandth of the mass of the lightest atom—hydrogen. More precise measurement later showed that it is even smaller, about one eighteen-hundredth as massive. Yet it had as much negative electrical charge as that atom might carry in positive charge.

It wasn't long before Thomson and others recognized that electrons were contained within all atoms. Furthermore, they were involved not only in electrical phenomena but also in the relationships between electricity and chemistry—including valence. Since atoms are electrically neutral, they must contain an amount of positive electricity to balance the negatively charged electrons, and that positive charge had to carry most of the atom's mass. But what was that

Particles Inside the Atom

Subatomic Pioneers. J. J. Thomson *(left)* discovered the electron, the first known subatomic particle, in 1897. Ernest Rutherford *(right)* discovered that the rest of the atom's mass is contained in a minuscule nucleus.

positively charged subatomic matter, and what was the internal structure of atoms?

J. J. Thomson suggested a model of atoms that resembled a popular British dessert, plum pudding, with tiny electron plums scattered throughout a positively charged bulk. Thomson's plum pudding atoms seemed sensible, but scientific models must be tested, and Ernest Rutherford (1871–1937) came up

The Neutron

with a way. He would probe the inside of atoms with radioactive beams. Rutherford had left his native New Zealand in 1895 to study the recently discovered phenomenon of radioactivity at Cambridge in Thomson's Cavendish Laboratory. By the time he left to become a professor at McGill University in Montreal, Canada, in 1898, he had discovered that radiation comes in two distinct forms. He named them alpha rays and beta rays after the first two letters of the Greek alphabet. At McGill, he and student Frederick Soddy discovered a third form of radioactivity in 1902, which they designated gamma rays. They also discovered that alpha radiation was a stream of energetic positively charged particles, while beta radiation consisted of high-speed negatively charged particles.

In 1907, Rutherford returned to England, becoming a professor at the University of Manchester and bursting with ideas about how to use radioactive beams. He would begin by shooting alpha particles through thin metallic foils. The way they would deflect, or scatter, might reveal the arrangement of the atoms in

Particles Inside the Atom

the foil—their size and spacing, perhaps even their shape. He and his student Hans Geiger developed an instrument to detect and count the alphas. They also demonstrated, as Rutherford had suspected, that alpha particles were helium atoms without their electrons.

By 1909, they were ready to begin alpha-scattering experiments. Nearly all the alphas passed straight through the foil or deflected only slightly. That result fit Thomson's model, except for one puzzling result. The counters were very accurate, and a few alpha particles were missing. Rutherford and Geiger had measured alpha scattering in every direction that seemed likely, but now they had to consider unlikely directions far off to the side. Intrigued but not wanting to divert Geiger from his main task, Rutherford turned to Ernest Marsden, a young student just learning the techniques of research. Marsden found the missing alpha particles. Astonishingly, not only had some alphas scattered far to the left or right of the original detector positions, but a few had even scattered backward!

The Neutron

What did the result mean? Rutherford realized that the atom was very different from what anyone had imagined up to that point. In 1911, he explained his results with a new atomic model. He envisioned atoms as miniature solar systems held together by opposite electric charges, which attract each other. He called the positively charged central body the nucleus (plural: nuclei) and noted that his results showed that the positive charge was concentrated in a region about one ten-thousandth of the size of the whole atom. The rest of the atom was empty space, except for the lightweight electrons, which he visualized as being like planets in orbit around that minuscule but massive nuclear sun. That structure explained why most alpha particles would pass through the foil with little scattering; they rarely came close enough to a nucleus to be deflected very much. Only on those rare occasions when a fast-moving alpha particle made a nearly direct hit on a much heavier nucleus would it scatter, and then it would be jolted so much that a sideways or even backward deflection would be possible.

Particles Inside the Atom

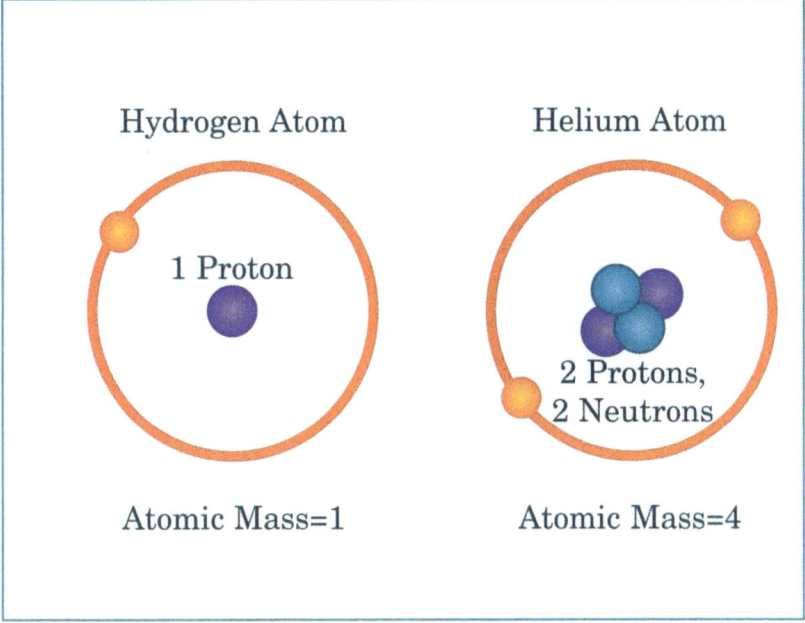

Too Much Mass. Once the atomic nucleus was discovered, scientists faced the question of what made nuclei of one element different from another. If they were made just of protons, then what would account for the difference between atomic number (the number of protons) and the atomic mass? Helium atoms, for example, had an atomic number of two but an atomic mass of four. Rutherford proposed that the extra mass came from particles that carried about the same mass as protons but were electrically neutral.

Weighty Problems

The nuclear model of atoms was quickly accepted, but that immediately raised new questions. For example, what was in the nucleus? Rutherford did more experiments

The Neutron

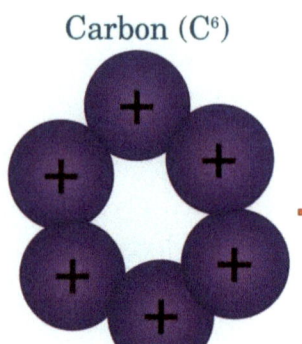

Carbon (C^6)

Carbon (C_{12}^6)

Neutrons and protons are bound by the strong nuclear force.

Nuclear Glue. Rutherford thought his neutral particles might also solve a problem with the force that holds the nucleus together. A powerful attractive force was needed to keep the electrical repulsion between positively charged protons from blowing the nucleus apart. Could interactions among protons and those neutral particles produce that force?

Particles Inside the Atom

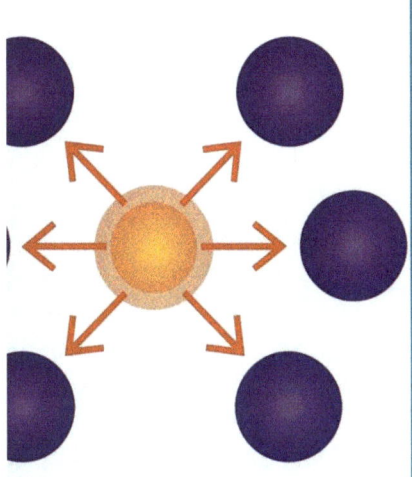

Electrical repulsion of protons would destroy the nucleus.

with alpha particles, and he became convinced that the hydrogen nucleus was a basic subatomic particle. He first called those nuclei "H particles," but he renamed them protons in 1917 when he detected them following alpha bombardment of boron, fluorine, sodium, aluminum, phosphorus, and nitrogen. That discovery changed scientific thinking about the meaning of the atomic number, which was at first defined as the number of an atom's electrons. Since it was not difficult to turn most atoms into ions by adding or taking away an electron or two, the atomic number was redefined as the total positive charge in

25

The Neutron

the nucleus. Hydrogen, the simplest atom with atomic number 1, has a nucleus with a single proton. Helium, with atomic number 2, has two protons, and so forth.

But things are not quite that simple when atomic weight—or atomic mass, a term that physicists prefer—is added to the picture. The atomic mass of hydrogen is one, and the hydrogen nucleus has one proton. But the helium nucleus—the alpha particle—which has two protons, carries an atomic mass of four. The discrepancy gets worse as atomic numbers increase. For atomic number 82, lead, the atomic mass is approximately 207. (The atomic masses turn out not to be exact whole numbers, and that will be explained later.) Protons do not account for even half the mass of most nuclei. Are there more protons, or might other subatomic particles exist?

Rutherford believed that the extra mass also has something to do with another difficulty in the planetary model. Electric force increases very rapidly as the separation of charged bodies decreases. Cutting the separation in half multiplies the force by four

Particles Inside the Atom

(2 x 2). At one-third the separation, the force is nine times as great (3 x 3). Since the nucleus is about one ten-thousandth the size of an atom and since two positive or two negative electric charges repel each other, two protons in the nucleus would push apart with a force millions of times as great as the attractive force between a proton and an orbiting electron! Such powerful forces would surely blow the nucleus apart—unless there was a stronger force within the nucleus to hold it together. Rutherford realized that whatever gives the nucleus more mass is probably also responsible for holding it together. The neutron, Rutherford's next important subatomic idea, turns out to be a very big part of the story.

Chapter Three

The Nucleus, Neutrons, and Nuclear Forces

No one was better prepared to discover the nucleus than Rutherford, thanks to his studies of radioactivity. Each radioactive element would produce identifiable alpha, beta, and gamma rays, which Rutherford and his student Frederick Soddy learned to distinguish. Soon they understood that radioactivity accompanied the kind of changes that alchemists had sought but were unable to achieve transforming one element into another.

Rutherford and Soddy were particularly interested in tracking the elements from their original form to their new forms. Since they had not yet discovered the nucleus, they did not yet know that the chemical changes they were observing were the result of nuclear events. But we can now describe the changes in terms of atomic mass and atomic number. For instance, when an atom emits an alpha

The Nucleus, Neutrons, and Nuclear Forces

particle, its atomic mass decreases by four units and its atomic number decreases by two. Likewise, the emission of a beta particle increases atomic number by one unit but doesn't change atomic mass. Both emissions result in "transmutation," the transformation of an atom of a "parent" element into an atom of a different element, the "daughter." The new atom is frequently more radioactive than its predecessor, so there is a chain of alpha or beta decays from one atom to another to another, and so forth. A gamma ray is pure energy, with neither mass nor electric charge, so the emission of a gamma does not cause transmutation. However, a radioactive atom emits a gamma ray only following some alpha or beta emissions, so a gamma is a sign that transmutation has recently occurred.

Rutherford and Soddy studied different radioactive decay chains, and they discovered in some cases that the daughter atoms in different chains behaved the same way chemically, which meant that they were the same elements. Yet they found that the two chemicals had different atomic masses. Soddy called such atoms with the same chemical behavior but different atomic

The Neutron

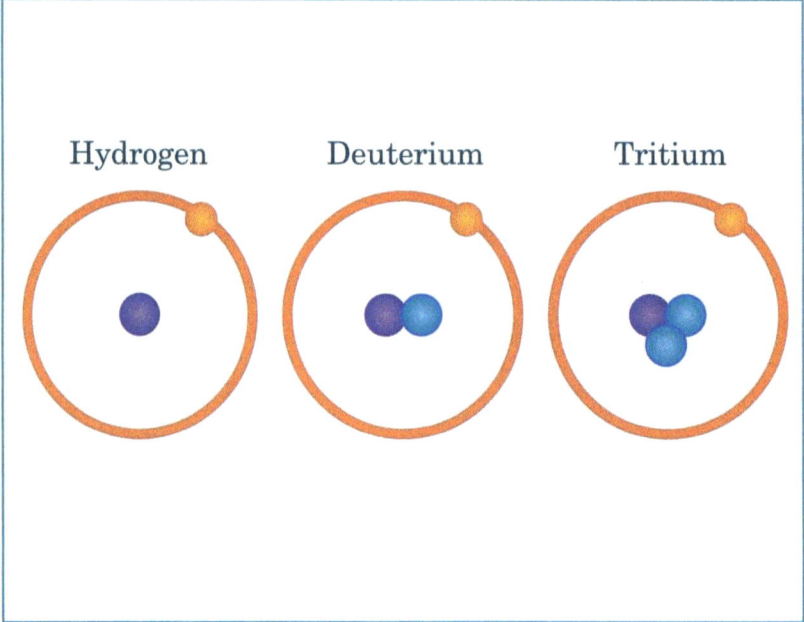

The Same, Only Different. In their study of radioactive decay chains, Rutherford and Soddy discovered that some atoms of an element had different masses than other atoms of the same element. They called such atoms isotopes. Years later, the difference between isotopes was understood to be a difference in the number of neutrons in the nucleus, and different isotopes were found to exist even for the lightest atoms. This illustration shows the isotopes of hydrogen, with atomic masses 1, 2, and 3.

masses isotopes, and he quickly realized that many nonradioactive atoms also had more than one isotopic form as well. To distinguish one isotope from another, scientists began denoting atoms by their chemical symbol and their atomic mass. For instance, natural chlorine is a mixture of two isotopes, Cl35 and the less common Cl37, with an average atomic mass of 35.47.

Discovery of the Neutron

By 1920, the year after Rutherford had replaced the retiring J. J. Thomson as leader of the Cavendish Laboratory at Cambridge University, no one questioned that the hydrogen nucleus was a single proton or that alpha particles were helium nuclei with two basic units of positive electric charge and an atomic mass of four units. But there were still questions about that extra mass that didn't come from protons. Some scientists suggested it was due to extra protons plus an equal number of electrons. Rutherford disagreed, explaining it this way: Because the nucleus is so small, any electron inside it would experience such a powerful electrical attraction to any proton that the pair would immediately bind together. The result would be an electrically neutral subatomic particle he called a neutron. Alpha particles, for example, consisted of two protons, each with one unit of positive electric charge, and two uncharged neutrons.

The Neutron

Rutherford now explained that alpha decay occurs when a stable helium nucleus—a unit of two protons and two neutrons—bursts out of a larger unstable nucleus. Likewise, he claimed that beta emission results from the splitting of a neutron in an unstable nucleus, resulting in a proton and an electron. Since electrons are so light, the new nucleus would have approximately the same atomic mass, but the added proton would increase its atomic number by one. He was right about alpha emission and almost right about beta emission. Later research showed that every beta particle has a tiny partner, a subatomic sprite called a neutrino with no electric charge and so little mass that no one has succeeded in measuring it.

Of course, Rutherford knew that any theory, no matter how good, still requires proof. He set out to find neutrons, and that was no easy task. During the 1920s, a number of scientists developed instruments that enabled them to see the paths of subatomic particles. These devices depended on the interactions between the subatomic particles and matter, especially their ability to ionize gases that they passed through.

The Nucleus, Neutrons, and Nuclear Forces

That worked fine for charged particles like protons and alphas, but not neutrons.

Finally in 1932, James Chadwick (1891–1974), one of Rutherford's colleagues at the Cavendish Laboratory, figured out a way to detect neutrons indirectly but convincingly. In 1930, two German researchers, Walther Bothe and Herbert Becker, discovered that bombarding beryllium metal, which had four protons and five neutrons in the nuclei of its naturally occurring isotope Be9, produced a powerful beam of neutral radiation. They assumed these beams were gamma rays because of the ease with which they penetrated matter. Following up, Irene Joliot-Curie, the

Identifying the Invisible. James Chadwick suspected that a strange beam of neutral radiation produced by other scientists probably consisted of never-before-detected neutrons, even though the other scientists believed the beam was composed of gamma rays. He then devised experiments to test his theory, which turned out to be correct. Scientists at last had convincing evidence that neutrons existed.

The Neutron

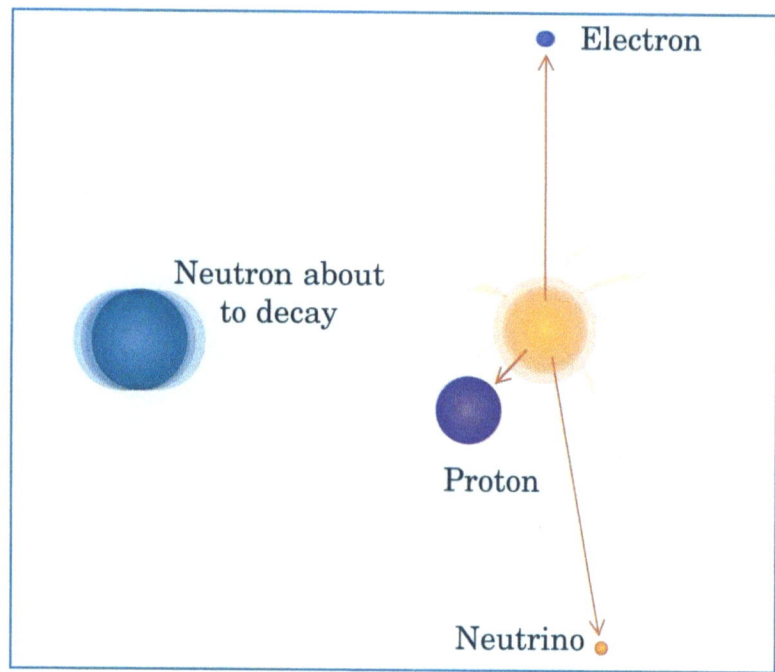

Neutrons and Beta Rays. The discovery of the neutron led Rutherford to describe beta emission as the result of a neutron decaying into a proton and an electron, the beta particle. Later research showed that this was not quite correct. Each beta particle is emitted with a tiny neutral partner called a neutrino.

daughter of the famous Pierre and Marie Curie, and her husband, Frederic Joliot, discovered that the neutral radiation would knock protons out of paraffin wax, which is rich in hydrogen. That was a surprising result for gamma rays, which could knock light electrons loose but had never been observed to eject heavier particles such as protons.

The Nucleus, Neutrons, and Nuclear Forces

When Chadwick heard of that result, he knew right away that the neutral beam had to be composed of neutrons. He did a series of experiments in which he allowed the beam to collide with a variety of gases. By measuring the scattering of the nuclei of those gas atoms, he was able to measure the mass of the particles in the beam, which turned out to be almost exactly the same mass as a proton, just as Rutherford had predicted for neutrons. That result established the basic atomic structure we now know: a tiny but massive nucleus of positively charged protons and electrically neutral neutrons, occupying only about a ten-thousandth of the atom's diameter, surrounded by light electrons in equal number to the protons.

Relativity, Quantum Mechanics, and Nuclear Forces

At the time Rutherford was probing the atomic nucleus, two other revolutions in physics, quantum mechanics and relativity, were also under way. Albert Einstein (1879–1955) had a hand in both. You have probably heard about

The Neutron

Einstein's famous equation $E=mc^2$. This equation comes from his 1905 work on the theory of relativity and expresses the unexpected idea that mass (m) and energy (E) are two aspects of the same thing. Since they are measured in different units, we need a conversion factor to match them up, just as you might change a measurement in inches to centimeters by multiplying by 2.54. To convert mass to energy, you multiply by the speed of light (c) times itself (or squared).

The power of that simple equation shows up in radioactive decay. When a radioactive nucleus emits an alpha particle, you might expect the sum of the mass of the alpha particle and the mass of the daughter nucleus to add up to the mass of the parent nucleus. But that is not true. The total mass after the decay is less than the original mass. The missing mass, when converted to energy by Einstein's famous equation, is exactly the amount of energy carried by the alpha particle.

Einstein solved another puzzle in 1905. The puzzle concerned the photoelectric effect, in which light could knock electrons free from metals—but only if its color was far enough toward the ultraviolet end of the spectrum. The

The Nucleus, Neutrons, and Nuclear Forces

color of the light was a measure of the frequency of light waves. That frequency had to reach a threshold before the light freed electrons. Einstein recognized a similarity between that threshold and an odd idea developed a few years earlier by Max Planck in his calculations of the spectrum produced by a hot body. Planck's equation depended on having light energy coming not in smooth waves like water, but in a stream of packets called quanta. Planck didn't believe that quanta actually existed, but they made his calculations work. Einstein's breakthrough was to recognize that the photoelectric effect was evidence that light quanta—which he named photons—were real.

That was the beginning of a new field called quantum mechanics, which states that all subatomic particles are quanta, and, just like photons, they act like waves under certain circumstances and particles at other times. For example, electrons in atoms have certain allowed wavelike states, each corresponding to a certain energy level specified by four "quantum numbers." The quantum numbers create a pattern as atoms increase in atomic number. Certain numbers of electrons act like

The Neutron

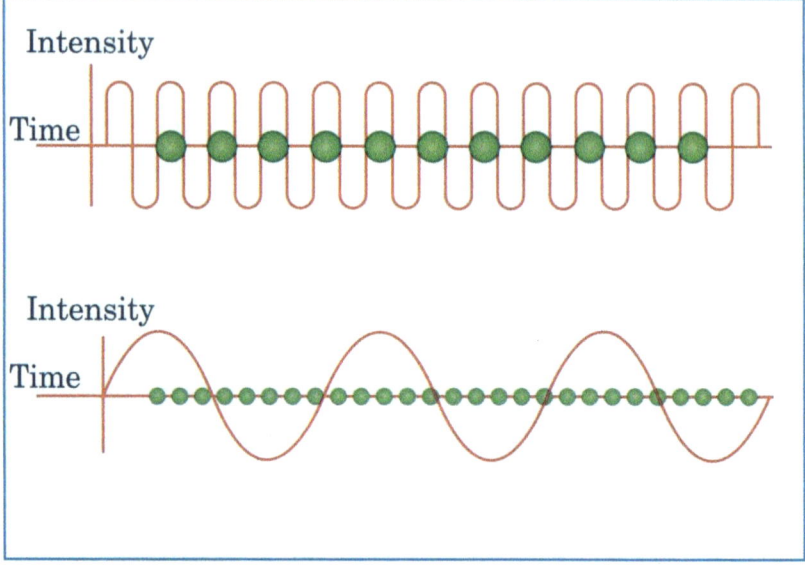

Powerful Packets. Albert Einstein's analysis of the photoelectric effect demonstrated that what Max Planck considered a mathematical trick—having light energy come in discrete packets called quanta (soon renamed photons)—was a physical reality. This sketch shows two beams of light with the same intensity (brightness) and different colors (frequencies). The high-frequency light has its peaks closer together and consists of larger photons than the low-frequency light. One large packet has enough energy to knock electrons loose from a metal, but the smaller packets, no matter how numerous, never break an electron free.

filled "shells," while the remaining electrons are available for atoms to interact. The pattern is the same one that Mendeleyev discovered. The table of elements is periodic because of quantum mechanics!

Quantum mechanics has been a spectacularly successful theory. Quantum mechanics changed

The Nucleus, Neutrons, and Nuclear Forces

the way physicists looked at the basic forces of nature. Applying the laws of electromagnetism at the atomic level required a new mathematical approach called quantum electrodynamics, in which attraction and repulsion are the result of an interchange of photons between electrically charged quanta, such as electrons and protons—and that takes us back into the nucleus. Since protons repel one another by exchanging photons, what keeps the nucleus from blowing itself apart? Another force must act inside the nucleus, and it must have something to do with neutrons.

That force is called the strong nuclear force, or simply the strong force. (Another nuclear force is called the weak force, and it explains beta decay.) The strong force has unusual properties compared to electromagnetism. For example, despite its power within the nucleus, the strong force must have a short range, quickly becoming weaker than electromagnetic forces as you move away from the nucleus. Otherwise, nuclei of different atoms would be drawn together and the universe would be one giant atom. Yet, within the nucleus, there must be a limit to the power of the strong force when

The Neutron

particles get too close. Otherwise, nuclear matter would crush itself to nothingness.

In the quantum world, nuclear forces can be explained by a theory called quantum chromodynamics. It is similar to quantum electrodynamics, with a few differences. "Chromo" refers to the Greek word for "color," but it has nothing to do with colors of light. Instead, physicists have borrowed the word to describe a property that nucleons—protons and neutrons—have through which they interact with the strong force, just as electric charge is the property that enables particles to interact electromagnetically. Instead of trading photons, nucleons attract each other by exchanging quanta called pi mesons, which have a mass of about 250 times that of an electron.

The strong nuclear force is just what nature ordered to hold nuclei together, but imagine what happens if a nucleus is too large for that force to be fully effective. It might release a lot of nuclear energy in a short time. The amazing thing is that people have figured out ways to harness that energy, and that is the main topic of the final chapter of this book.

Chapter Four

Neutrons and Nuclear Fission

In the quantum world, nothing stands still. You might think of radioactive decay in this way. A nucleus is made up of nucleons (protons and neutrons) that are constantly rearranging themselves. Like electrons in atoms, protons and neutrons have states within the nucleus described by quantum numbers. Also like electrons, those states have certain closed shells that are particularly stable. For instance, an alpha particle has two protons and two neutrons filling the lowest energy shells for each particle. If four such nucleons in a nucleus join as an alpha particle, it may break loose from the rest of the nucleus. That happens more readily as the nucleus gets large enough that the strong force begins to drop off at its outermost parts.

Beta radiation is a similar phenomenon involving the weak force. A free neutron outside a nucleus is unstable because it

The Neutron

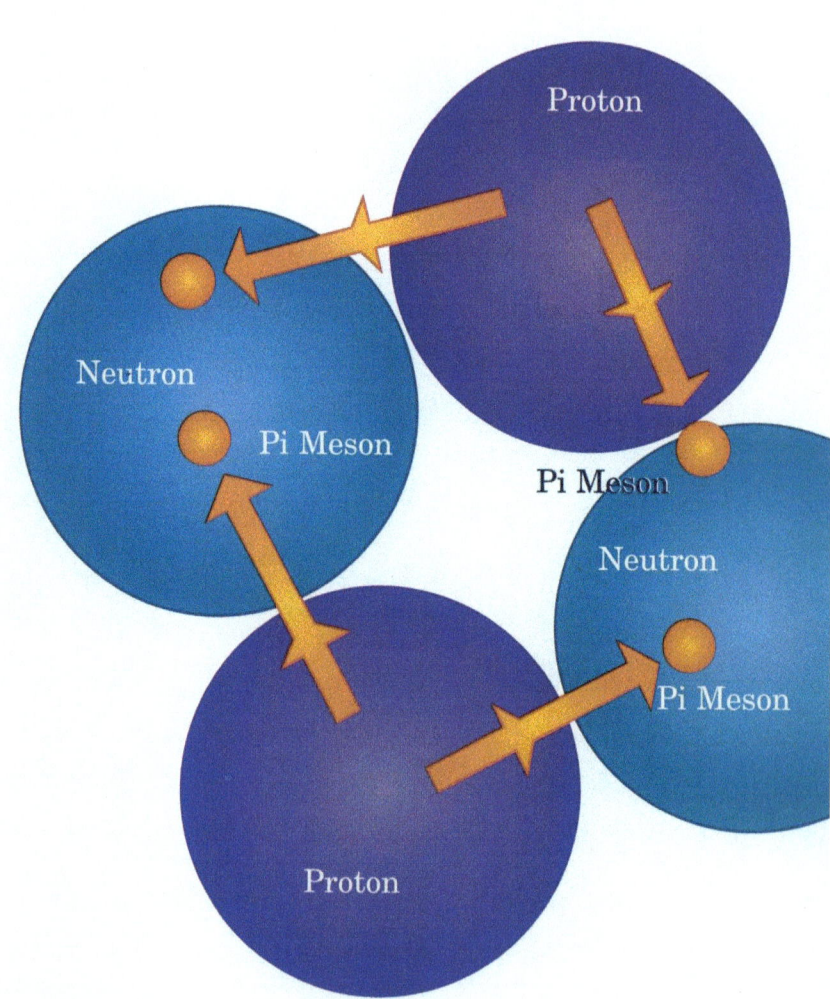

The Strong Nuclear Force. Inside many nuclei, neutrons do not experience beta decay because they are always changing identities with protons by exchanging pairs of "virtual" pi mesons. The pi particles flicker into and out of existence as permitted by the uncertainty principle of quantum mechanics. The exchange of those particles results in an attraction between protons and neutrons. That attraction, the strong nuclear force, overcomes the electrical repulsion between protons and holds the nucleus together.

Neutrons and Nuclear Fission

has more mass than a proton, electron, and neutrino combined. It is always on the verge of blowing itself apart into those particles—which is what happens in beta decay—except for the weak nuclear force that holds it together. If you could watch such a neutron for about fifteen minutes, it would have a 50-50 chance of decaying. Inside most nuclei, the neutrons stay together because of quantum chromodynamic effects. The protons and neutrons are always passing pi mesons back and forth and changing from one type of particle to the other, and as a result, the neutrons don't have enough time to decay.

The Neutron

Those pi mesons are called virtual particles because they have much more mass than that given up by the protons and neutrons that release them. They thus use energy that should not be available. But a rule of quantum mechanics, called the uncertainty principle, allows them to flicker in and out of existence long enough to travel the short distance between the nucleons. That also explains the short range of the strong nuclear force since the particles that carry it disappear too fast for it to have a long range. In larger nuclei, neutrons stay neutrons a little longer, and sometimes the weak force is not enough to prevent beta decay.

Fission

Before the strong and weak forces were understood, scientists had learned enough about nuclei to understand what was happening inside the nucleus when alpha, beta, and gamma rays were emitted. In 1938, two German physicists, Austrian-born Lise Meitner (1878–1968) and Otto Hahn (1879–1968),

Neutrons and Nuclear Fission

Bits and Pieces. After some experiments he did in 1938 with Fritz Strassman, German scientist Otto Hahn described the odd results to his Jewish former colleague, Lise Meitner, who fled to Sweden to escape the Nazis. Uranium nuclei seemed to be splitting approximately in half. Meitner and her nephew Otto Frisch quickly developed a theory that explained the results, which became known as nuclear fission.

began investigating artificial elements beyond uranium (atomic number 92) in the periodic table. These had first been noticed by Italian physicist Enrico Fermi in 1935 in experiments in which he bombarded uranium with neutrons. A uranium nucleus (U238) would capture a neutron and quickly emit a beta,

The Neutron

leaving behind a proton in the new nucleus with atomic number 93 and atomic mass 239 (polonium 239). The unstable Po239 nucleus would emit another beta, leaving behind a nucleus with atomic number 94 and atomic mass 239, plutonium 239 (Pu239).

At the same time in Paris, Irene Curie-Joliot and Pavel Savitch had noticed the odd presence in the bombarded uranium of an element that behaved chemically like the much lighter lanthanum (atomic number 57). About to flee Germany for Sweden because of the danger to Jews like herself from the Nazi government, Meitner discussed the odd results with Hahn and Fritz Strassman, who then tried experiments similar to those of Joliot-Curie and Savitch and found barium (atomic number 56). It was as if uranium nuclei were splitting into two parts!

In Sweden, Meitner enjoyed a year-end visit from her nephew Otto Frisch, also a physicist. They discussed the odd results and figured out what was happening. A large nucleus might behave like a water droplet bouncing back and forth into an elongated hourglass shape and

Neutrons and Nuclear Fission

back again. By chance, the distribution of protons and neutrons between the two parts might be just right to form two separate smaller nuclei within the larger one, and the electrical force between the two pieces would blast them apart in a process called nuclear fission. Furthermore, because larger nuclei tended to have a larger proportion of neutrons than smaller nuclei, there would be a few neutrons left over. Fission would be even more likely if the nucleus was in an excited state, such as would occur when it is struck by or captures a neutron.

Nuclear Bombs and Nuclear Power

It didn't take long for scientists all over the world to figure out what might be possible. Each fission produces some energy plus more neutrons coming out of the nucleus than the one absorbed by the nucleus. If those outgoing neutrons could cause other fissions, the entire piece of matter could quickly be engulfed in a chain reaction, producing

The Neutron

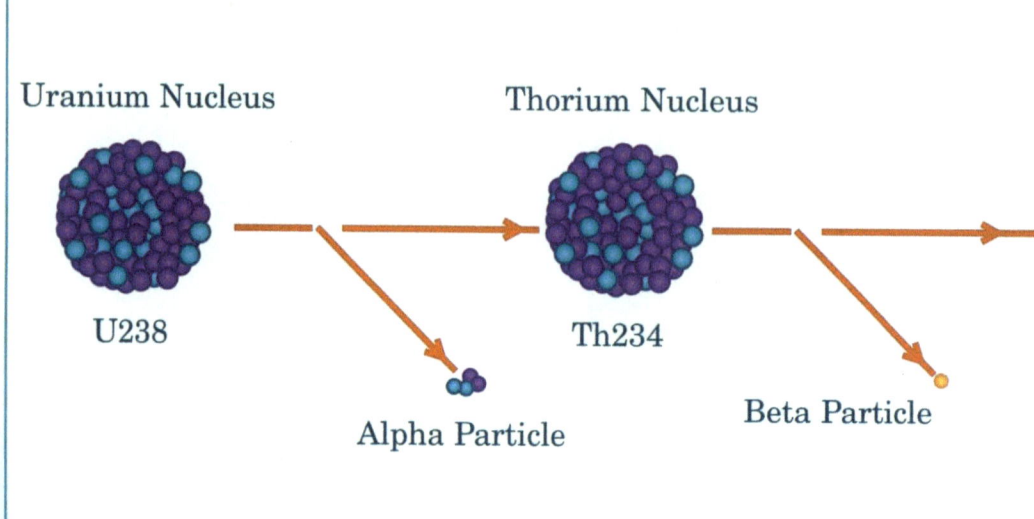

From Uranium to Lead. Rutherford and Soddy's measurements showed that one nuclear decay was often followed by another and another, eventually leading to a stable nucleus. They traced several different decay chains, some of which resulted, after several steps and billions of years, in the transmutation of uranium into lead.

enormous energy from a relatively small amount of mass, according to Einstein's famous formula $E=mc^2$. With World War II looming, governments on both sides saw the potential to apply fission for military use. They envisioned bombs with frightful power, and they quickly set to work.

Neutrons and Nuclear Fission

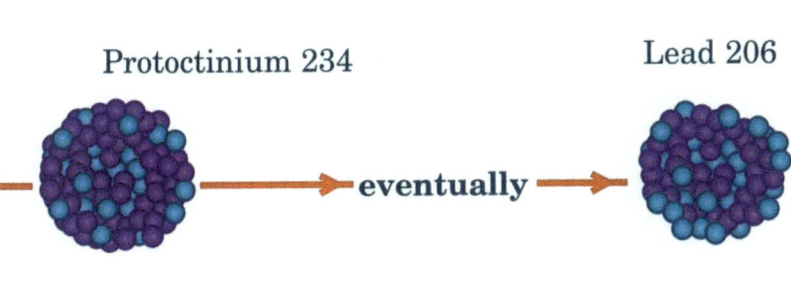

Making a uranium bomb is not as easy as it might seem. It's easy for neutrons to escape. If the average fission event produces three neutrons, then a chain reaction would require that one-third of the neutrons cause another fission. One problem is that most uranium nuclei are U238, which can capture neutrons but do not decay by fission. Less than 1 percent of uranium exists as the "fissile" U235. To make a uranium bomb, the first step is to enrich the uranium, increasing the fraction of

The Neutron

Fission as a Weapon. Nuclear fission produces high-energy neutrons that can strike other nuclei and cause them to break apart as well. If conditions are right, fission can produce an explosive chain reaction—a powerful bomb. The artificial isotope plutonium-239 can produce a more powerful blast from less material than uranium-235, as was demonstrated in this deadly plutonium fission bomb explosion over Nagasaki, Japan, at the end of World War II. After the war, many nuclear scientists and engineers devoted their efforts to producing electricity from controlled nuclear fission.

U235 atoms. An explosion requires a "critical mass" so that not too many of the neutrons escape the material before interacting with a fissile nucleus, and that critical mass must be

Neutrons and Nuclear Fission

brought together quickly and stay together after "igniting" the explosion.

The remaining "depleted" uranium can be used to make bombs in another way. As Fermi had observed, U238 can capture a neutron and, after two beta decays, become Pu239, which turns out to be more fissile than U235. As you probably know, the United States ended World War II when it used two fission bombs in Japan. What you may not know is that the first bomb used uranium and the second used plutonium.

After the war ended, many engineers turned their efforts to building electrical power plants that get their energy from nuclear fission. But they needed to find a design that would control the chain reaction and—most important—automatically shut it down in case of trouble. The key to their designs was the fact that most neutrons produced in fission reactions move so fast that they zip out of the material. They would be much more likely to cause a fission reaction if they pass a nucleus slowly enough for it to divert them from their path. So

The Neutron

nuclear reactors were built with cores of fuel rods separated by a "moderator"—water in most cases, but some use graphite—that slows the neutrons enough to cause fission. They also have control rods with materials that absorb neutrons and stop a chain reaction. These must be withdrawn from the core against powerful springs in order to start the chain reaction. In an emergency, or if a spring fails, the control rod snaps into the core and stops the reaction.

Nuclear power plants have excellent safety records, but there have been frightening failures or near failures. They also produce waste that is dangerously radioactive and must be safely disposed of. The future of nuclear power in the United States is uncertain, but many countries rely on it for much of their electricity. It is safe to say that neutrons will be responsible for much of the world's electricity for a long time to come.

Glossary

alchemy A predecessor field to chemistry through which many people hoped to transform less valuable metals into gold but never succeeded.

alpha particle or alpha ray A helium nucleus that is emitted from some radioactive elements.

atom The smallest bit of matter that can be identified as a certain chemical element.

atomic mass or atomic weight The mass of a particular atom compared to a standard, which sets the mass of a carbon-12 atom to be exactly 12. For a particular isotope, that value is approximately the number of protons plus the number of neutrons in its nucleus. For a naturally occurring element, that value is approximately the number of protons plus the average number of neutrons in the nuclei of naturally occurring isotopes.

atomic number The number of protons in the nucleus of an atom, which determines its chemical identity as an element.

beta particle or beta ray An electron that is emitted from some radioactive elements.

The Neutron

chain reaction A process in which neutrons produced in the fission of one nucleus cause fission in other nuclei, which produce more neutrons and more fissions until the available fissile material is largely consumed.

color In the theory of the strong nuclear force, this term is used to refer to the property that particles have that makes them respond to the force, just as electric charge is a property of particles that makes them respond to electromagnetic forces.

compound A substance made of only one kind of molecule that consists of more than one kind of atom. For example, water is made of molecules that contain two atoms of hydrogen and one atom of oxygen.

electromagnetism A fundamental force of nature, or property of matter and energy, that includes electricity, magnetism, and electromagnetic waves, such as light.

electron A very light subatomic particle (the first to be discovered) that carries negative charge and is responsible for chemical properties of matter.

Glossary

element A substance made of only one kind of atom.

emission Sending out something that has been produced, such as the emission of an alpha, beta, or gamma ray from a radioactive atom.

fission A form of nuclear decay in which a nucleus splits into two smaller nuclei plus a few neutrons. Nuclei that decay in this way are called fissile.

gamma ray A high energy photon that is emitted from some radioactive elements.

molecule The smallest bit of matter that can be identified as a certain chemical compound.

neutron A subatomic particle with neutral electric charge found in the nucleus of atoms.

nucleus The very tiny positively charged central part of an atom that carries most of its mass.

periodic table of the elements An arrangement of the elements in rows and columns by increasing atomic number, first proposed by Dmitry Mendeleyev, in which elements in the same column have similar chemical properties.

The Neutron

photoelectric effect A phenomenon in which light can, under some circumstances, knock electrons out of atoms. Einstein's explanation of this effect led to scientific acceptance of the photon as a particle and eventually to quantum mechanics.

photon A particle that carries electromagnetic energy, such as light energy.

pi meson A particle that is interchanged by nucleons (protons and electrons) to produce the strong nuclear force

proton A subatomic particle with positive electric charge found in the nucleus of atoms.

quantum electrodynamics and quantum chromodynamics Quantum mechanical formulations that express the electromagnetic and strong nuclear interactions between particles.

quantum mechanics A field of physics developed to describe the relationships between matter and energy that accounts for the dual wave-particle nature of both.

radioactivity A property of unstable atoms that causes them to emit alpha, beta, or gamma rays or to undergo fission.

Glossary

scattering An experimental technique used to detect the shape or properties of an unseen object by observing how other objects deflect from it.

spectrum (pl. spectra) The mixture of colors contained within a beam of light, or the band produced when those colors are spread out by a prism or other device that separates the colors from each other.

strong nuclear force or strong force A fundamental force of nature that acts to hold the protons and neutrons in a nucleus together.

theory of relativity A theory developed by Albert Einstein that deals with the relationship between space and time. Its most famous equation ($E=mc^2$) described the relationship between mass and energy.

transmutation The transformation of one element to another by a change in its nucleus, such as by alpha or beta emission.

weak nuclear force or weak force A fundamental force of nature that is responsible for beta decay of a radioactive nucleus.

For More Information

Organizations

Lederman Science Center
Fermilab MS 777
Box 500
Batavia, IL 60510
Web site: http://www-ed.fnal.gov/ed_lsc.html
This museum is an outstanding place to discover the science and history of subatomic particles. It is located at the Fermi National Accelerator Laboratory (Fermilab) outside of Chicago.

Magazines

American Scientist
P.O. Box 13975
Research Triangle Park, NC 27709-3975
Web site: http://www.americanscientist.org

New Scientist (U.S. offices of British magazine)
275 Washington Street, Suite 290
Newton, MA 02458
Web site: http://www.newscientist.com

For More Information

Science News
1719 N Street NW
Washington, DC 20036
Web site: http://www.sciencenews.org

Scientific American
415 Madison Avenue
New York, NY 10017
Web site: http://www.sciam.com

Web Sites

Due to the changing nature of Internet links, the Rosen Publishing Group, Inc., has developed an online list of Web sites related to the subject of this book. This site is updated regularly. Please use this link to access the list:

http://www.rosenlinks.com/lsap/neutron

For Further Reading

Bortz, Fred. *Catastrophe! Great Engineering Failure—and Success*. New York: W. H. Freeman, 1995.

Close, Frank, Michael Marten, and Christine Sutton. *The Particle Odyssey: A Journey to the Heart of Matter*. New York: Oxford University Press, 2002.

Cooper, Christopher. *Matter* (Eyewitness Books). New York: Dorling Kindersley, Inc., 2000.

Henderson, Harry, and Lisa Yount. *The Scientific Revolution*. San Diego: Lucent Books, 1996.

Narins, Brigham, ed. *Notable Scientists from 1900 to the Present*. Farmington Hills, MI: The Gale Group, 2001.

Pringle, Laurence. *Nuclear Energy: Troubled Past, Uncertain Future*. New York: Macmillan, 1989.

Strathern, Paul. *Mendeleyev's Dream: The Quest for the Elements*. New York: Berkeley, 2002.

Bibliography

Close, Frank, Michael Marten, and Christine Sutton. *The Particle Odyssey: A Journey to the Heart of Matter.* New York: Oxford University Press, 2002.

Cropper, William H. *Great Physicists: The Life and Times of Leading Physicists from Galileo to Hawking.* New York: Oxford University Press, 2001.

Nobel Foundation. *Nobel Lectures in Physics 1901-1921.* River Edge, NJ: World Scientific Publishing Company, 1998.

Strathern, Paul. *Mendeleyev's Dream: The Quest for the Elements.* New York: Berkeley, 2002.

Young, Hugh D., and Roger A. Freedman. *University Physics: Extended Version with Modern Physics.* Reading, MA: Addison-Wesley Publishing Co., 2000.

Index

A
alchemy, 8, 9, 28
alpha rays/radiation/particles, 20, 21, 22, 25, 26, 28–29, 31, 32, 33, 36, 41, 44
Aristotle, 7, 8
atomic number, 25, 26, 28, 29, 32, 37, 45, 46
atomic mass/weight, 11, 12, 15, 16, 18, 26, 28, 29, 30, 31, 32, 46
atoms, 4, 5, 6, 7, 8, 9, 10, 12, 15, 18, 20–21, 27, 28–29, 30, 37, 38, 39, 41, 50
   combinations of, 8–9, 11
   "daughters," 29, 36
   "parents," 29, 36
   properties of, 5, 10–11, 17, 25, 29
   structure of, 19, 22, 35, 37–38

B
beta rays/radiation/particle, 20, 28, 29, 32, 39, 41, 43, 44, 45, 46, 51
bombs, 48, 49–51

C
Cambridge University, 17, 20, 31
Cavendish Laboratory, 20, 31, 33
Chadwick, James, 33, 35
chain reaction, 47–48, 49
chemistry/chemists, 8, 9, 10, 11, 12, 18
compounds, 5, 8, 9, 12

D
Dalton, John, 4, 5, 9–11
   atomic theory of, 4, 5
Democritus, 4, 5, 7, 8

E
Einstein, Albert, 35–37, 48
electric force, 26–27, 47
electricity, 9, 17, 18, 40, 52
electrodes, 17
electromagnetism, 39, 40
electrons, 5, 17, 18, 19, 21, 22, 25, 27, 31, 32, 34, 35, 36, 37, 38, 39, 40, 41, 43
elements, 4, 5, 7, 8, 9, 11, 12, 15, 16, 17, 28, 29, 45
   classification of, 15–16
   properties of, 12, 16, 17, 29–30
energy, 18, 29, 37, 36, 40, 47, 48
England/English, 4, 9, 17, 20

F
Fermi, Enrico, 45, 51
fissile, 49, 50

G
gamma rays, 20, 28, 29, 33, 34, 44
gases, 9, 32, 35
Geiger, Hans, 21
Greeks, 4, 6–7, 8, 17, 20

H
Hahn, Otto, 44–45, 46
helium, 21, 26, 31, 32
hydrogen, 4, 8, 11, 18, 25, 26, 31, 34

I
ions, 25, 32
isotopes, 30, 33

J
Joliot, Frederic, 34
Joliot-Curie, Irene, 33–34, 46

L
Leucippus, 4, 5, 8
light, 36, 37

M
Marsden, Ernest, 21
matter, 5, 6, 7–8, 9, 32, 47
   composition of, 4, 6, 10
Meitner, Lise, 44–45, 46
Mendeleyev, Dmitry Ivanovich, 12–15, 16, 38
molecules, 4, 5, 8, 11, 12

Index

N
negative electrical charge, 18, 20, 27
neutral charge, 18, 31, 32
neutrino, 32, 43
neutrons, 5, 16, 27, 31, 32, 33, 35, 40, 41, 43, 44, 45, 47, 49, 50, 51, 52
nuclear energy, 40, 51, 52
nuclear fission, 47, 49, 51, 52
nuclear reactors, 51–52
nucleons, 40, 41, 44
nucleus, 22, 23, 25, 26, 27, 28, 31, 32, 35, 36, 39–40, 41, 43, 44, 45, 46–47, 49, 50, 51

P
periodic table of the elements, 16, 17, 38, 45
photoelectric effect, 36
photons, 37, 39, 40
physics/physicists, 5, 35, 39, 45, 46
pi mesons, 40, 43, 44
planetary model of atom, 22, 23, 26
plum pudding atomic model, 19
positive electrical charge, 18–20, 22, 25, 27, 31, 33
protons, 5, 25, 26, 27, 31, 32, 33, 34, 35, 39, 40, 41, 43, 44, 46, 47

Q
quanta, 37, 39, 40
quantum chromodynamics, 40, 43
quantum electrodynamics, 39, 40
quantum mechanics, 35, 37, 38–39, 44
quantum numbers, 37, 41

R
radioactivity, 20, 28, 29, 36, 41, 52
relativity, 35, 36
Rutherford, Ernest, 19–22, 23–25, 26, 27, 28, 29, 31–32, 33, 35
atomic model of, 22, 26

S
scattering, 21, 22, 35
scientific observation, 4, 6, 9
shells, 38, 41
Soddy, Frederick, 20, 28, 29–30,
spectrum, 36, 37
strong nuclear force, 39–40, 41, 44
subatomic particles, 5, 8, 17, 19, 25, 26, 27, 31, 32, 37, 43, 44

T
Thomson, J. J., 17, 18, 20, 21, 31
atomic model of, 19
transmutation, 29

U
uranium, 45, 46, 49–50, 51

V
valence, 15, 16, 18

W
weak nuclear force, 39, 40, 41, 43, 44
World War II, 48, 51

About the Author

Award-winning children's author Fred Bortz spent the first twenty-five years of his working career as a physicist, gaining experience in fields as varied as nuclear reactor design, automobile engine control systems, and science education. He earned his PhD at Carnegie-Mellon University, where he also worked in several research groups from 1979 through 1994. He has been a full-time writer since 1996.

Photo Credits

Cover, pp. 1, 3, 14–15, 23, 24–25, 30, 34, 38, 42–43, 48–49 by Thomas Forget; pp. 7, 10 (right), 19, 33 © Science Photo Library; pp. 10 (left), 50 © Bettmann/Corbis; p. 13 © Steve Raymer/Corbis; p. 45 © Photo Researchers, Inc.

Designer: Thomas Forget; Editor: Jake Goldberg

www.ingramcontent.com/pod-product-compliance
Lightning Source LLC
Chambersburg PA
CBHW041115070526
44584CB00002B/174